HOME SELLING SECRETS

Raphael Orozco Jr.

SalesGeex

The author reserves all the right to this book. They do not permit anyone to reproduce or transmit any part of this book through any means or form be it electronic or mechanical. No one has the right to store the information herein in a retrieval system, or to photocopy, record copies, scan parts, etc., without the proper permission of the publisher or author.

All Right Reserved

Copyright © Raphael Orozco Jr., 2020

Description

This book is a short summary of the modern-day strategies a home seller can implement in the sale of a home. It is particularly based on strategies implemented by real estate marketing agency XOLDE.COM

Disclaimer

All the information in this document is to be used for informational and educational purposes only. The author will not account in any way for any results that stem from the use of the information herein. While conscious and creative attempts have been made to ensure that all information provided herein is as accurate and useful as possible, the author is not legally bound to be responsible for any damage caused by the accuracy as well as use/misuse of this information.

Contact

For more information or to contact the author please go to www.xolde.com or visit @0rozcojr

Table of Contents

INTRODUCTION 6
CHAPTER 1 16
    THE THREAD 16

CHAPTER 2 21
    TRAFFIC 21

CHAPTER 3 30
    KNOW YOUR MARKET 30

CHAPTER 4 36
    USEFUL DATA 36

CHAPTER 5 53
    AVERAGE DAILY USE 53

CHAPTER 6 62
    PROFESSIONAL PHOTOGRAPHY 62

CHAPTER 7 72
    HOME STAGING 72

CHAPTER 8 77

PROFESSIONAL VIDEO	77
CHAPTER 9	**84**
THE HOMEBUYER PSYCHOLOGY	84
CHAPTER 10	**95**
PRICING	95
CONCLUSION	112

INTRODUCTION

Tom just got a new job offer in a different state. He has to sell his house and move. Jack and Jill have a new kid. They realized that the place they got after they were married wouldn't be adequate to raise a family, so they are looking for buyers. The Smiths are beginning to get uncomfortable with how commercial and noisy the neighborhood is getting. They want to move somewhere quiet and peaceful. They also want to sell their home. A young bachelor decides that he can make do with a smaller place. He decides to sell and pocket the profits. All these people have a couple of things in common. The first is that they want to sell their homes. The second is that they want to make as much profit as possible.

As expected, home sellers want to clear out their sales fast, and they want to come out of it with as much money as possible. Although, a lot of the time, this might not happen. There is a tendency to focus on what you want to get out of it that you pay less attention to how it is done. Three things might happen in this case. The first is that the house does not sell. The second is that it sells late, and the third is that it sells below the market value.

Homeowners can all relate to the euphoria of buying a house. You're proud of yourself, you're happy, and you're excited. You have it at the back of your mind that value is going to appreciate over the years. It feels good to know that. It feels good to know that all your hard work will keep yielding over time. One thing that is not evident at the time is the fact that you could jeopardize your possibilities of getting a great deal if you don't take care to learn the right way and employ it.

Nobody wants to see their hard work and diligence go for naught. There is a particular type of energy that people use to protect what is theirs, especially when it comes to specific investments. It might not have been so hard for some people to own a home, but for others, it would have required a greater amount of effort. What this book seeks to achieve is to help you protect and preserve what is yours. The truth of the matter is that you not only want to sell your home fast, but your ultimate goal is to sell for more money.

There's a tiny problem. Your listing agent doesn't work for free. And you will be tempted to hire a discount brokerage to help you sell.

We've all been in a position where people have taken advantage of us. One way or another. Can you think of a time someone took advantage of you over money? How did that make you feel?

It doesn't feel good. We don't like to be hustled. And we don't like being lied to. It's one of the worst insults someone one can give us. And it happens in any business, any industry.

It's easy to ignore that historically, homeowners have been willing to pay large amounts of money to sell their home. That's because real estate is one of the two major areas where people can make hundreds of thousands of dollars by simply buying and holding, same as with stocks.

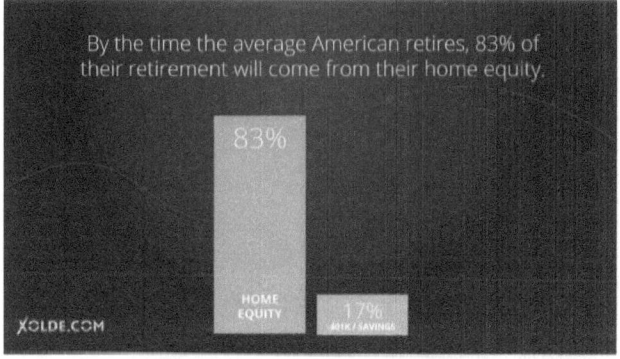

So, when it comes time to sell; people are willing to pay a real professional whatever it takes to get the most out of their sale instead of trying to do it themselves and losing money.

They say money is the root of all evil. Are you familiar with that quote? It's a very popular quote from the bible, but it's actually incomplete. The real bible actually says, "The <u>love</u> of money is the root of all evil" (1 Timothy 6:10). I'm no bible expert but I'm pretty sure it refers to greed.

You want more money for your house, and I'm here to tell you that it's okay. Wanting more money is okay. Yes, you can and yes you should want more money. We all want money because money makes our life easier. Money equals freedom.

But what about the real estate agent? Should you be worried? Of course, you should, there are unethical people in any profession. But your biggest concern when choosing a listing agent is not what you're thinking of. The answer lies in the word value. We can all make more money as long as we start with one simple principle. **Provide value.**

You don't have to be sleezy to make more money. You don't have to screw people over. You just have to give value. The only reason someone like you and I wouldn't think we can get paid what we want is because we don't value what we bring to a transaction. We don't feel we're worth it.

One doesn't simply make a lot of money just because he wants to. We live in a value driven society. We return reciprocity based on value given. People pay money to have problems go away. You make money because you solve problems. Because you provide value. Right? Value is compensated with money. And we get paid in proportion to the difficulty of the problems you solve. Period. And so, the goal of money is to reward/motivate people for adding value to our society. Which leads to your biggest area of concern, your listing agent is happy to take your money, but does he provide you with enough value?

You're about to enter most important transactions of your life. To some, the sale of your biggest asset; your home. You are entering into a contract to pay a given commission fee for the sale of your home. But how do you know if you're getting the right value in return? At this point, you don't know. You have no have control. If you're the type of person that likes to have full control of your life and your future, then this book is for you.

You're going to learn a new system that would help homeowners like Tom, Jack and Jill, and the Smiths sell their homes faster and for more money. It is working for numerous sellers out there, and you're next. You too can learn how sell that house without sacrificing your equity.

There is something that I'm going to be referring to as "on the market" time. As it sounds, it means the time that a house spends on the market before it is sold. Over the course of this book, you are going to learn how to save yourself a lot of "on the market" time. We are talking months. Not only that, but you are also going to learn how to prevent losing thousands of dollars in the sale. This loss could be a result of wrong marketing or selling below value.

By now, you must have come across home sale platforms like Zillow, Trulia, Redfin, Realtor.com, and others. These are some of the most renowned real estate platforms consumers rely on. You are going to see the secret to capturing the attention of those home buyers that are checking out your competition, and you are going to learn how you can lure them your way. You will be shown how to make your home one of the top-performing houses in your market on real estate platforms like Zillow. You are going to do all of this without having to adjust your price week after week.

Think of this as a game. Imagine real estate agents as the players. Compared to you, they are veterans. One thing all veterans have in common is experience. There are tricks you don't know about; little methods used by real estate agents to sell homes. They use these tricks on home sellers, and it could cost the seller a lot of money. You're going to learn all about it.

By the time you get to the end of this book, you would have learned about the market. You would have been privy to advanced tricks and tactics used in the market; today's market. You would have learned how to get yourself the best possible deal.

You would have learned how to do it fast, and you would have learned how to do it without having to lower your expectations and sacrifice your equity. What's more, you would have learned how to do it all while shirking the stress and difficulty that could have otherwise been a factor.

"Give me six hours to chop down a tree, and I'll spend the first four sharpening the ax." - Abraham Lincoln.

The objective is to ready yourself. Take the time to read this book and gain the required knowledge to sell your home profitably. Take the time to "sharpen your ax" and learn how to spot real value.

CHAPTER 1

THE THREAD

Selling a home isn't like selling a book at a garage sale. It definitely isn't like selling a car. The fact is, selling a house is a different ball game, and we all know this. Most sellers are aware of the simple truth that it probably won't be easy. In this awareness, precisely, lies the crux.

As I said, most home sellers are aware that it won't be easy, but I'm going to liken that awareness to a thread. Imagine you have a piece of clothing that has thread sticking out. I'm sure we have all had this at least once before. I'm sure, also, that we have tried to pull the thread. If you're lucky, it cuts, and your cloth looks as good as new once again. More often than not, it doesn't cut. It keeps pulling and pulling, messing up the clothing. This would continue until you give up and get a pair of scissors.

The awareness of the problem is just like the analogy; there is more to it. Most homeowners know of the difficulty in selling a home, but they are not aware of to what degree. They know it is not easy, but they are still majorly ignorant of the main obstacle ahead. This obstacle is the relationship between demand. Which is clearly reflected in the amount of mortgage applications.

Major Obstacle

- Mortgage Applications
- Demand Goes Down
- And Supply Goes UP

- **HUGE PROBLEM**

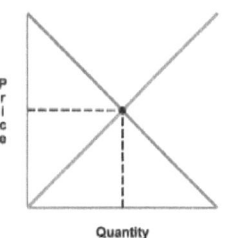

If you paid attention in economics class, you would understand the relationship between demand and supply. Regardless, I will go over it once again. The way it works is that there is a direct relationship between demand and supply. When demand goes up, supply goes down. When supply goes down and demand is up, we would have low supply: a sort of scarcity in the market. What this does is that it causes the price of whatever is being sold to rise above the ordinary. For the sellers, they would be making more than they did before from the same quantity of product.

When demand goes down, supply goes up. When this happens, we would have a high supply, and this is bad for business. Just like the scarcity creates an increase in price when supply goes down, in this situation, the opposite happens. The increase in supply, as opposed to demand, causes the prices to fall. The sellers make less than they did from the same quantity of product. Think of it like a seesaw; when one goes up, the other comes down.

The 2020 pandemic immediately had a massive impact on the economy. It doesn't take a specialist to realize this. The demand for a lot of things went down drastically, and this also affected supply. In real estate, it caused less people to apply for mortgages, and when it comes to prospective home buyers, the number of people actually getting approved for a loan took a hit due to banks being overly cautious and implementing conservative guideline modifications.

In real estate, a home buyer demand is characterized by home buyers who are "Ready, Willing and Able." This means home buyer demand is only affected by those who are actually capable of making a home purchase, whether it be through a home financing or cash. Keep in mind that the uncertainty preceding a recession will already be a negative factor affecting demand. And when you add the fact that banks are pre-approving less home buyers, the situation worsens. And just like a seesaw, as I have explained, low demand naturally causes supply to go up.

The reason for the venture into the economics lesson is that when demand is down, prices are down. We have already established that what you want is the best possible bargain for your house and this poses a huge problem.

Less Home-Buyer Traffic

- High Supply and Low Demand = PROBLEM

TRAFFIC

No Traffic = No Sales = No Profit = Failure

All this mumbo jumbo has an effect on your traffic. I'm going to establish exactly what that means and how it affects you before going further.

CHAPTER 2

TRAFFIC

The make-or-break factor in any transaction. As the name already suggests, traffic is like the attention on your home. The importance of homebuyer traffic for your sale cannot, in honesty, be overemphasized. It is the make or break factor in the whole shebang. Your homebuyer traffic is directly related to your marketing. The better your marketing, the better your deal and the less you market a deal... you get the gist.

I have made it known that low demand means high supply. Low demand also means less buyer traffic. You already know how vital homebuyer traffic is, so you can make a quick inference about how much of a problem this will be, especially in this period.

Without having the right amount of traffic, you can't make a good sale, and **everything** is about making that sale. I want you to imagine yourself as the owner of a business. You need the traffic to make sales, right? It is the same with this. The business owner's profit is tied directly to his consumer traffic. The more the traffic, the more he earns, and the less the traffic, the less he/she earns. If a business struggles with traffic, that means it would be struggling with getting clients, and, ultimately, nobody wants that. That is the situation for a business owner. For you, a home seller, the absence of significant traffic means you don't have exposure for your home. There is no visibility, and this means there won't be a successful sale. The thought of that should leave a bitter taste in your mouth.

The point of this exposition is to help you understand how important homebuyer traffic is. According to Zillow, you need more than 280 online visits within the first week to be able to sell in less than 30 days.

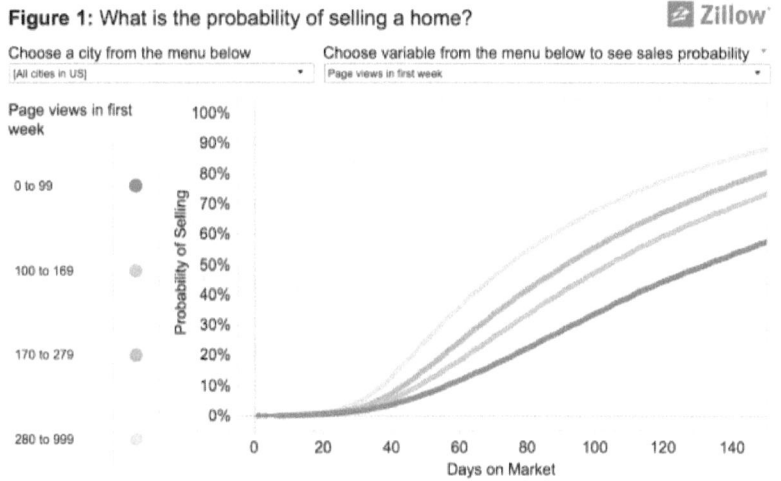

Figure 1: What is the probability of selling a home?

If you hire a licensed real estate agent to oversee the sale, you generally would not have to worry about the cost of marketing. As customary, the way it works is that if the home sells, you pay a percentage of the sales price to the agent. This means that prior to the sale, you don't have to worry about any of the costs relating to the marketing campaign. Isn't that great?

I imagine that at this point, you are wondering about how much you will have to pay for selling your home. Traditionally, the seller pays about 6% of the sales price in commissions. It can be higher, depending on the investment required on behalf of the listing agent. This percentage is then divided between your agent (the listing agent), and the buyer's agent. It is also customary for the commissions to be split right down the middle. Meaning, the listing agent would expect 3%, and the buyer's agent would expect 3%.

You are the seller; the boss. You are paying the big bucks. You deserve to know where your money is going and how it is being invested. If you are making the assumption that the listing agent would use a budget of this 3% to facilitate the marketing, you would be wrong. It is a bit more complicated than that.

A real estate agent's take-home commission is a bit of a private topic and sellers rarely openly inquire about it due to common courtesy. But you will now have a chance to get a better understanding of what's a bit of a mystery. Here is an example of what the sale of a million-dollar home could look like under traditional structures. Take careful note of the numbers.

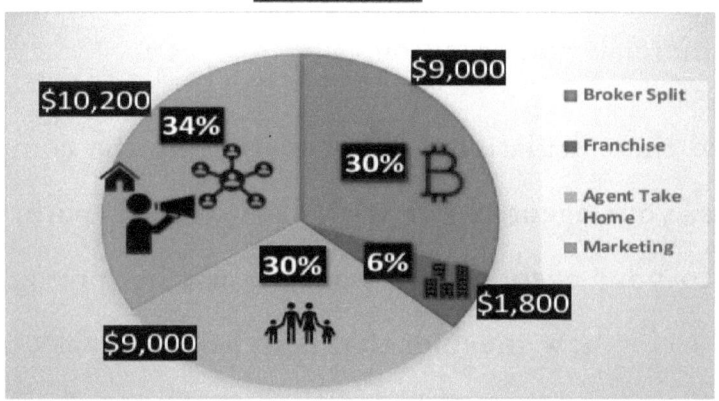

The fact is this: most real estate companies operate in a structure very similar to what is above. In the industry, it is what is referred to as the 70/30 split. You might have heard that before, you might not have. What it means is that 70% goes to the agent, and the remainder (30%) goes to the brokerage.

Take a look at the structure; the part you have in red labeled as broker's 30% split is the brokerage's share. A little bit for sustainability; to keep the lights running, etc. In addition to that, brokerages usually tack on a 6% franchise fee. This extra fee is justified by a company's marketing campaigns. It's how they pay for commercials and branding.

Make the calculations. After everything the company takes, your agent is left with 64% to work with. It is left to your agent at this point to use his prerogative and decide how much of that 64% he would take home and how much he would put into marketing the home. Now, if your agent is someone that is very considerate and fair, he/she would use a large chunk of that 64% to market the house, and he would take home the rest, let's say 34% to market and 30% to take home.

You have to consider your real estate agent became an agent to earn a living. He/she likely has a family to support. And for an agent, as for any other professional, their objective is to make money and so they do not like to gamble with their funds. They would be putting their money on the line. Remember that they only get paid if the home sells. All the investments into marketing are funded from the agent's own pocket until the sale is made. Because of this, the agent is likely to invest only a small amount of his potential commission into marketing the home. This is most likely going to be the low cost and traditional forms of real estate marketing. Those which, generally, do not generate much home buyer traffic.

Now that you have a good picture of how it works with these companies, it would interest you to know that there are alternatives. There are some companies out there that operate differently. These companies allow their agents to part with the totality of their commission. That's right, the whole 100%.

You might be wondering how they would be able to make their profits if they do that. In a nutshell, the way it works is that the real estate brokerage charges the agent a monthly office fee and a small flat fee per transaction completed.

So, if you are paying thirty thousand dollars for the sale of your million-dollar home, the agent would expect 100% of it, minus his monthly office dues and a small transaction fee. A complete game changer. For the agent, this is much cheaper than having to pay 34% as it is in the case of a traditional brokerage. This is not only more beneficial for the agent, but this works better for you because you would have an agent who is able to use a more substantial marketing budget to invest in helping you sell your home. A larger marketing budget increases the potential to generate more traffic to your home.

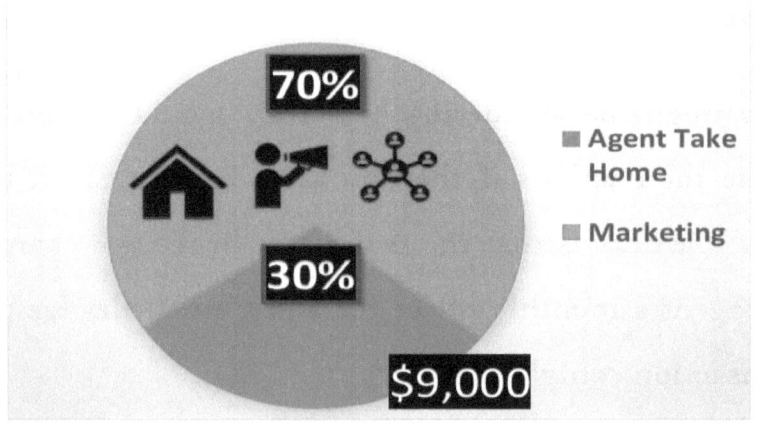

As the real estate market adjusted and technology has changed over the years, savvy real estate agents are not only learning modern marketing campaigns and how to run them, but they are teaming up with brokerages that provide the support needed to better fund and invest in effective marketing campaigns that will benefit not only the seller, but the real estate market as a whole.

Moving on from the realm of brokerage, the next thing for you to understand is that you have to know your market. You're now going to be learning how to take your market out on date night and get all the little details.

CHAPTER 3

KNOW YOUR MARKET

Your "market" refers to the whole enchilada. But in this case, it is not referring to the prices of homes in the area. It is referring to the home buyers in particular. You have to understand them.

By now, you know how relevant traffic is. As a home seller, what you want now is exposure. Guess what, everybody does. The home sellers say that they want more exposure on their home. The real estate brokers swear that they want more exposure. Even real estate companies want more exposure. Exposure is like the dark ring from Lord of the rings; everybody wants it.

Everybody says the same thing: they want more exposure. The reason for this is simple. When you get more exposure, it means you are getting more attention. When you are getting more attention, you are getting more money. Money is everybody's motivation here. For that house you want to sell, you need all the attention in the world. You want to get as much exposure as you can. Granted, you want someone that can show you the ropes. You want someone to walk you through the whole process of selling your home, but what you want more than anything is that exposure. You want more exposure to generate the homebuyer traffic you need because when you get the homebuyer traffic that you need, it produces the buyers.

Take a moment to ask yourself. What is the first thing that comes to your mind when you think of marketing exposure?

I can bet that most people would say Open Houses. To a lot of people, Open Houses equal marketing exposure. The truth is, agents do that more for their self-promotion than otherwise. It is the best way they know how to do this. In fact, it is one of the only ways. I mentioned earlier that the pandemic has affected the economy. That has also stretched to this. Due to COVID-19, most MLS's (Multiple Listing Service) regulations immediately prohibited open houses indefinitely, and you can imagine the effect of that.

It caused an immediate shift in the way that traditional agents marketed themselves. In the long run, it couldn't have much of an impact on sales because open houses were never a reliable way to sell homes in the first place.

I told you that you need to understand your market. Now it's time to hit you with some facts and figures. The National Association of Realtors does research periodically to help keep track of trends. Basically, they want to see what is working in today's market.

In 1981, 22 percent of home buyers used newspaper ads to find homes. The latest research NAR conducted shows that currently about 93% of home buyers search for their homes online.

It shouldn't be surprising. Everyone has a smartphone now, and hardly would you find anyone who is not active on social media. The internet and social media came like thieves in the night and completely disrupted industries. The real estate industry is not left out. This change is not a bad thing. The disruption came in a good way. It has created an entirely new method of marketing. It brought with it, digital marketing.

Once again, I am going to remind you that it is important to treat the sale of your home like it is a business; like it is a business seeking customers. You need to understand though, that as a home seller, you are not in the business of selling real estate. You are really in the data business. You are in the business of leveraging consumer data that is essential to stealing traffic from the other active homes in your neighborhood and directing it towards your own home. You want to steal that attention and make it yours.

As a home seller, you should be stealing away customers from other home listings competing with yours. The word "steal" must have made you uncomfortable, as it should. Stealing is unethical. But don't worry, there is actually a completely ethical way to steal, and the way you can do it is through leveraging online consumer data (see what I mean? It's the reason you're actually in the data business).

CHAPTER 4

USEFUL DATA

These are some valuable online data points that digital marketers can use to target homebuyers:

- Credit Score
- Income
- Inclination to Buy
- Mortgage Loan Applicants
- Homeowners vs Renters
- Home Address
- Work Address
- Legal Status

Credit Score

Focusing your marketing campaign targeting on individuals with higher credit scores will immediately hyper target homebuyers who are most likely in a position to successfully qualify for a home loan and weed out those who would not. If you can avoid wasting marketing dollars on consumers whose internet data points indicate they would not be a good candidate to buy your house, you could potentially double your marketing efforts on a higher quality consumer.

Income

Your focus is not just random traffic. You also want to reach the people that can buy your house. Certain houses fit a specific type of people in terms of wants, but this also applies in terms of income.

Like credit score, debt to income ratios are also big factors in a home buyer's ability to qualify for a home loan. And a lender will most definitely evaluate an aspiring home buyer's income to determine the size of the loan he could potentially qualify for. If you are selling a one million dollar home, advertising your home to homebuyers whose income bracket would likely place them in a much lower price range would not make much sense, would it? Higher-income earners are more likely to qualify for the higher-priced homes, whereas low-income earners are most likely limited to the lower-priced homes. Your listing agent and his lender colleagues should have a general idea of the ideal income range that your model home buyer "Persona" should have.

Inclination to Buy

An inclination is a disposition to do something. It is just as it sounds. Certain potential buyers in the market are more inclined to buy than others.

Artificial intelligence and premium algorithms are the true makeup of real estate technology. And you can leverage this to predict the likelihood of a consumer not only responding to the marketing of your home, but also the likelihood of them actually taking action to purchase your home. This inclination could be as a result of a range of factors, including but not limited to the type of house the homebuyer is looking for, the location, their income, size of the family, etc. Whatever it may be, inclination to buy ratings prove there are buyers more motivated than others and a digital marketing platform can focus on targeting.

Mortgage Loan Applications

According to a Consumer Housing Trends Report conducted by Zillow in 2018, about 77 percent of home buyers finance their home. This means a large majority of home buyers must apply for mortgage prior to purchasing. Ironically, most home buyers start shopping for a home prior to even speaking to a mortgage loan professional and seeking a pre-approval letter. What this means is that aspiring home buyers with the highest inclination to pull the trigger on a home would have already visited internet sites seeking a mortgage application. A valuable data point indicative of a home buyer's very high inclination to buy.

Homeowners vs Renters

Homeownership is another data point available for advanced targeting. And as it relates to real estate sales, it would be significant to realize that renters who are actively looking to purchase a home would most likely fall under the category of first-time home buyers. The budget or purchase power of a first-time homebuyer will generally be considerably lower that of homeowner seeking his second home.

Here's why: Statistically, a first-time home buyer will generally seek a low down payment mortgage loan. Which is pretty much any loan requiring a down payment below the traditional 20%. For obvious reasons, a homeowner seeking a mortgage for their second home will most likely have enough equity in his home to qualify for a traditional mortgage requiring a down payment larger than 20% of the loan.

Why is this important? Well, low down payment mortgage loan programs have a maximum loan limit. These limits vary by county to county and are based on your county's median home prices.

Home Address

If and when data and research show a considerable amount of people from another region in the country are moving into your region. In-bound and out-bound mover's data can help effectively target the marketing of your home to those people according to their current home address. For example, if you live in Los Angeles, and the latest trend indicates there is a large amount of in-bound traffic from Seattle, it would behoove you to assure your home is being marketed in Seattle.

You want to target buyers who want to live where your home is.

Work Address

Another category of home buyer would be those who are moving because of work. For obvious reasons, these people would want a property that' is close to their workplace. They may want a home that is generally close to the corporate scene. And these sorts of buyers are going to be particular about the location.

In this scenario, a viable option would be to target potential homebuyers who live outside your city but show interest in your city based on their recent online activity and documented interests.

Legal Status

Last but not least, legal status. In order to qualify for a mortgage loan in the United States, the borrower must be a legal resident. Just like people with bad credit scores, there would be very little benefit to marketing a home to an audience whose online data points indicate they lack legal residency and therefore would not be a candidate to purchase your home.

You want to aim for the right people and avoid foolishly wasting adspend (Money spent on advertising).

There is a lot of useful information out there which can be leveraged to market a home. When used properly, it is priceless. It is an apparent fact that you would benefit enormously from hyper-targeting the marketing of your home to a specific group of people, or in marketing terms; an "Audience."

Marketing is expensive. Real estate marketing is especially competitive and therefore costlier than other niches. As a home seller, you want as much exposure as possible, but it is expensive, and you are working on a budget. Remember you are only promising to pay your listing agent a small agreed upon fee, and after all, the goal is for you to make as much as possible from the sale of your home.

If there was a way to focus your home's marketing campaign on that particular group of people more likely to buy your home, wouldn't you want to use that?

How cool would that be? Being able to avoid beating around in the dark, and effectively focus your budget on the buyers with the most potential.

With how costly marketing is, think of how wasteful it would be to market your home to people who aren't even good candidates. Time, effort and money are all going down the drain. And pretty soon, your marketer's budget will run out.

With digital marketing, it is possible to focus your attention on the specific few who are most likely to purchase a product. In your case your home. This takes the phrase "target market" to a whole different level.

All buyers aren't the same. And your home shouldn't be marketed as if they are. Cookie cutter marketing plans don't cut it. Marketing is a process, not a checklist; it's a science, not an art.

A buyer for a $500,000 home is a different buyer than one for a $1,500,000 home. That's why you need a specific marketing plan to target a list of home buyer's specific to your home. Your home is unique and the marketing to sell it needs to be unique, too.

Using internet data points to create a "Buyer Persona" can tell you the type of person and family who will find your home ideal. With a buyer profile, we know what the buyer finds important and where the best place is to put your home in front of them. With this information, special messaging and media plans specific to your home can be created to target your buyers.

But do it the right way, or it can backfire on you. It should be noted that there are laws and regulations in place which prohibit certain types of audience targeting or audience exclusions in real estate advertising. The Fair Housing Act prohibits property owners, financial institutions, and landlords from discriminating against prospective tenants or buyers on the basis of race, religion, national origin, sex, family status, or disability in the United States. Serious penalties are in place for violating the Fair Housing Act, and it is crucial that you hire a professional who is not only aware of these rules in real estate advertising, but someone who will adhere by the rules.

At this juncture, we are going to take a little detour into Psychology. Ever heard of the Reticular Activating System?

Did you ever own binoculars as a kid? The way a binocular works is that it zooms in on a particular place or area, and blocks out everything else that is not on its radar. The Reticular Activating System works just like that. It blocks out unimportant content so that only what is necessary at that point in time can get through. It is a bunch of nerves in the brain.

You could also think of it like a search feature on a website. You type in specific keywords, and your device sifts through what is available, ignoring everything else but the thing you're searching. It is not so intentional or controlled in our minds, but that is, basically, how it works. It is our brain's idea of having a tunnel vision.

Have you ever noticed that sometimes, when you talk about something new; something you've never seen before, all of a sudden you begin to see it everywhere. For instance, you notice a particular car model, and after that day, it suddenly seems like you are seeing that model everywhere when you never used to see it before then. It could perhaps be that you remarked at how well the color yellow looked on a person's skin and you begin to notice that quite a lot of people seem to be wearing the color yellow from that point on.

Sometimes, it could be that someone in your family is pregnant. All of a sudden, you start running into pregnant women everywhere you go. It could even get so much that you begin to feel like there is a secret conspiracy going on in which all the pregnant women in the area hatched a plan to follow you around. Could it be that you're being stalked?

The truth is you're only experiencing the Reticular Activating System which works as a filter in our minds. It sieves through everything and zones in on certain things.

Here we are; the reason for the detour. What if I told you that there was a way you could tap into a similar filter for all the qualified home buyers that are currently in your area? That's right — a way to reach all the buyers who displayed interest in purchasing a home just like yours.

You probably have not realized it yet. There is something out there that is making you think of stuff that you want to buy; it is making you think of them over and over again, and it seems like a amazing coincidence. But it's actually quite simple. It's a concept called retargeting, and social media platforms allows for some of the best retargeting methods.

You might have noticed that whenever you do some online shopping, the products that you are checking out seem to follow you everywhere you go while you are online. Ads seem tailored to those specific products that you have been searching for. This is what retargeting is.

Retargeting is a digital marketer's version of the Reticular Activating System. For years, major brands have been making use of this to redirect you back to them. It is one of the methods they employ to drive you right back to their stores.

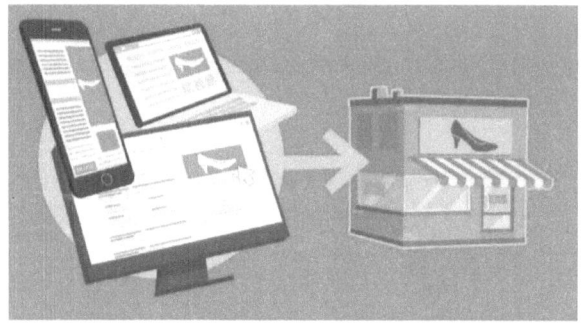

There is a straightforward way to achieve this for your home sale. The **secret** to doing this successfully would be to tag your home listing with a simple code like this one:

```
<script>
// Google Analytics tracker initialization snippet
(function(i,s,o,g,r,a,m){i['GoogleAnalyticsObject']=r;i[r]=i[r]||function(){
(i[r].q=i[r].q||[]).push(arguments)},i[r].l=1*new Date();a=s.createElement(o),
m=s.getElementsByTagName(o)[0];a.async=1;a.src=g;m.parentNode.insertBefore(a,m)
})(window,document,'script','https://www.google-analytics.com/analytics.js','ga');

// Creating 'owox' tracker that will duplicate original Google Analytics hits to Google BigQuery.
ga('create', 'UA-XXXXX-Y', 'auto', {'name':'owox'});
// Define Create Only fields here after you've created a tracker

// Connecting OWOX BI Streaming plugin to the tracker
ga('owox.require', 'OWOXBIStreaming', {sessionIdDimension: ZZ});

// OWOX BI Streaming plugin code
(function(){var h=function(f,b){var g=f.get("sendHitTask"),h=function(){function a(a,e){var d="XDomainRequest" in window?"XDomainRequest":"XMLHttpRequest",c=new window[d];c.open("POST",a,10);c.onprogress=function(){};c.ontimeout=function()
{};c.onerror=function(){};c.onload=function()
{};c.setRequestHeader&&c.setRequestHeader("Content-Type","text/plain");"XDomainRequest"==d?
setTimeout(function(){c.send(e)},0):c.send(e)}function g(a,e){var
d=document.createElement("img");d.onload=function(){};d.src=a+
```

Although it looks scary and difficult to the average person, this wouldn't be anything complicated for digital marketing experts. It would be as easy as copy and paste for them.

By adding this simple code to your home listing, your real estate agent would not only be able to track everyone who showed a level of interest in your home, but he would also have access to advanced software. This software allows for the ability to create similar audiences and duplicate those similar audiences repeatedly using the data points the audiences share in common. All these would just require the marketer to spend some time analyzing the data and selecting the right options. It really does work like magic. After all that has been done, all that is left is for your home to follow them all around social media like the stench of a strong cologne that you used too much of. In essence, that means activating their Reticular Activating System (pun unintended).

CHAPTER 5

AVERAGE DAILY USE

Do you know that the average person spends an average of 2 and a half hours every single day on social media? A whopping 2.5 hours! This should make it clear that if you're going to do any type of marketing for your house, digital marketing is a must. It is non-negotiable. Marketing through digital platforms is the only way to track the people who showed any interest in what you are selling; your very own home.

You might ask, but is this simple trick not being used by real estate agents?

The short answer is no, they don't know how. Contrary to popular belief, real estate agents mostly focus on their sales skills. They pay more attention to their people skills; interrelation, interaction, and communication skills. They don't invest enough time to improve their **marketing** skills as a home seller would expect. When an average person pictures real estate, he or she pictures open houses, and that is because a lot of listing agents are still stuck in the outdated traditional methods.

However, open minded of real estate agents are starting to come around. They are beginning to understand the value of digital marketing across social media platforms and are open to invest in it. They are hopping on the wave. Regardless of that, they still lack the understanding of the real power behind it. They haven't fully tapped into its potential. They've been leaving out the essential part: retargeting.

For those real estate agents that actually invest in doing marketing, the traditional and more popular way of real estate marketing is to send out postcards or deliver them in person. The problem with this is that there is no way to track who exactly is looking at their postcards and flyers with genuine interest and intent. And as I mentioned before, the only feasible way to do this is through digital marketing. Digital marketing is the only practical way anyone can track everything and everyone at a large scale.

Numbers. They say numbers don't lie. You and I could make a bold statement and argue that 82 percent of statistics are made up on the spot. And that ten out of seven people don't even understand fractions. Ten out of seven? Exactly. Truth be told, statistics could easily be manipulated to support any argument. For that reason, when evaluating statistical data it is not only important to consider the source but to evaluate the motivation behind their research and whether it was conducted impartially or biased. Nevertheless, here are some numbers that a home seller would find interesting and why should too:

Statistics

** Sources include the National Association of Realtors, Study.com and others.

- 7 Seconds
- 93% - Visual Appearance
- 70% - Sensory Receptors
- 48 Seconds
- 05:50 with video

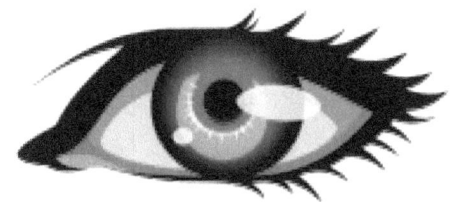

No matter how well targeted your advertisement seems to be, you still have to hit the mark. Aiming is only half the struggle; you also need to hit the bull's eye. You should have something in place to capture and keep people's attention.

While they are surfing, you need to pique their interest. You want your house to be that intriguing option they stop at. Otherwise, they would simply skip right over your home and move on to the next thing.

• Take a look at those **7 seconds** sitting above. It is not just a random number. The average attention span for a normal person is 7 seconds. This means that you have only seven seconds to play fisherman and catch their eye. This doesn't just apply to visuals. Whether or not you will command a person's attention, especially in various forms of presentation, all depends on the first few seconds. If you don't interest them right from the jump, it would be difficult to get their attention later on. If you don't catch the attention of your homebuyer in those first 7 seconds, they are gone.

• Now we are looking at that **93%** standing there. Have you ever been in a gathering and you meet certain people that just seem to catch your eye? Even before these people speak, it feels like they would be someone you are going to enjoy talking to. You start to wonder, and you think to yourself, "there is just something about them." We have all been there. I am willing to bet that there have also been situations where some things just appeal to you, and you can't seem to place a finger on it. The reason is simple. Humans are incredibly visual creatures. All that glitters is not gold, but we sure seem to love them nonetheless. This also applies to purchases. **93%** of homebuyers place visual appearance above all other factors when shopping. We live in a very visually stimulated world.

There is another angle that you might not have considered. You host visitors in your home. Both those that you can't stand and those you are seeking to impress. You must have found yourself in a situation at least once before where you are scrambling to clean up the house because you are expecting a visitor. Something about people is that they like to feel proud of what is theirs. We like to feel good about our clothes, our shoes and our cars. We **love** to feel good about our homes. To many, it's our most prized possession. It might seem like a normal part of everyday living, but it sounds good to be told, "Hey, I like your place." As a homeowner, I'm sure you are aware of this. Keeping this random piece of knowledge in mind, definitely, when buying a house, nobody would go for the less aesthetically pleasing option.

- Over the route of our expository journey together, you've been exposed to a bunch of random facts. From Economics to Psychology, you have learnt about them all. You are about to learn a little something else about anatomy this time. Sensory receptors are those things in our physical makeup that are responsible for receiving stimuli. What this means is that they perceive something and then communicate it to your brain. **70%** of sensory receptors exist in the eyes. This means you can't drop the ball with those visuals. There is no other way to go about it. Your listing **has to be** visually pleasing.

- As mentioned earlier, you have about 7 seconds to catch your buyer's attention. Once you do this, you have another **48 seconds** to dazzle them. According to real estate research, the average time a buyer spends on a real estate listing online is 48 seconds. When the marketing of your home includes video, that time increases to about **5 minutes and 50 seconds**. Think of it like window shopping. You're walking around, looking into different stores. Then you find something that attracts you. You walk in, check out what you want to, and you leave. Credit to that store for bringing you in. You continue with your exploring. Once again, something catches your eye. You walk into this store, but it is different this time. The service is excellent. The sales representatives are really friendly and helpful. They are patient with you. They answer all your questions. They even draw your attention to other things that you might be interested in. Now, you are spending all your time in this store. Chances are, they retain you as a long-term customer.

You can also achieve this with your home listing by simply adding a well edited video tour. In fact, it should be your goal. There is a whole lot that you can do to keep those buyers with you. How much more time do you think a buyer will spend on your listing if you add a 3D virtual tour? This gives buyers the option to navigate through your house and imagine themselves living there. When they see that, don't you think they'll remember your house above the rest?

CHAPTER 6

PROFESSIONAL PHOTOGRAPHY

Before   After

Which home is more attractive?

It is very possible that both of these pictures were taken with professional cameras. That is not all that is required to make an image stand out like that, though. A lot more goes into it. There is a lot of work that a professional photographer can do to enhance an image and make it pop like the picture in the second frame.

I will be exposing you to a bit about photography in this section and what goes into producing top-notch images.

You see a lot of memes on the internet from time to time. People take a couple of scenic pictures and suddenly start to label themselves as photographers. No doubt that those pictures would have been brilliant, but regardless of how good your eye is, a professional photograph is still a professional photograph.

You might have planned to take the pictures yourself. You have learnt a few things about appealing to the visual receptors of an average human, so you feel well equipped to handle it. By the time I go through this, you should be aptly informed about the things that go into getting a good picture.

First off, pictures are different. Photographers specialize in different areas because of this. Portraits are different from events coverage. There is nature photography, fashion photography, abstract photography, sports photography and a whole lot more. The point is that there are many types of genres out there. Even if you feel that take flawless pictures of your family members and you are the designated photographer at events, those are still very much different from this.

Taking shots of your home veers to the side of architectural photography. Before I go into what that is all about, I am going to establish some general things about photography and why you need professional help.

Number one, lighting is essential. Often, when people take pictures, they rarely pay attention to this. People take pictures backing the sun. They take pictures with shadows strewn across their faces. Often, photos are either overexposed or underexposed. This might not be an issue if you are posting it on your Facebook page, but you are trying to make a sale, so it is a big deal.

The timing is important. Pictures taken during the day are blessed with natural sunlight, and that is the best type of lighting there is. This is true, but it is also true that you need to try taking the pictures at different times. Depending on the home; its look, location and vibe, the images might actually come out looking better in the evening or night as opposed to the daytime. You are going to need to experiment and try out different things before you arrive at the right fit.

You need to take the pictures from the right place - hit the angles, and all that. You can't just walk outside with your phone and take one shot. That won't do it. Quality requires effort. Small buildings might be easier to capture. You probably won't have a lot of problem getting the building in all its cuteness. When it comes to large buildings, it could get a bit trickier. You want to capture the property in all its size and glory. You don't want to downsize its beauty or anything like that. Care has to be taken to ensure that you shoot from the right place.

Positioning the shot also mandates that you take the environment into cognizance. If the home has a nice-looking garden or lawn, you want to showcase all these. Something about flowers and well-kept greenery adds a lot of appeal to a picture. You don't want this to be lost on your buyers. Apart from this, the neighborhood and environment are also important. You want your home to be photographed in a way that highlights the environment around it well. It should fit perfectly with what is around it while also standing out. You must have heard some people say that they look either better or worse in pictures. Pictures could make something look better than it is. They could also make it look worse.

There is still a bit more to it than that. There are specific details and secrets that only someone with experience would have the privilege of knowing about. There is a little something called the rule of thirds. On every picture, there are imaginary lines that cut across the frame. There are two vertical and horizontal lines dividing the image into squares. You might have seen this on your phone camera while taking a picture. They help you position the shot the right way. It makes it easy to focus on precisely what it is that you want to focus on while relegating what's left to the background. There is **still** more to it than that, though. That grid also helps you to take a shot in a way that allows communication between the subject and its environment.

You could take a picture and have something smack in the center. All the focus is on it. That is good if it is what you are going for. You want your home to appropriately communicate with the environment and paint a wholesome picture for the viewer. It won't cut it just to point the camera and shoot.

In a picture, and on the grid, there are certain areas that the eyes unconsciously drift towards. You might not have noticed because it seems like you just take in whatever you can see. There is a science to it. Some pictures are extremely easy on the eyes, and others are not. There are some pictures that contain a lot, but it all seems like **one,** but there are some pictures that are just disorienting. You want your photos to come across as the former.

A professional photographer will place the essential things in those focus areas. Some of these focus areas are where the lines intersect. This is done while also letting the other stuff in the shot be seen at the same time. Your pictures would come out looking a whole lot better if there is an emphasis in the rule of thirds.

If you are thinking of simply stepping out there to take a couple of pictures with your iPhone and call it a day, I would advise that you kill those intentions. It doesn't matter how you want to play it, pictures taken with professional cameras are better than those that are taken with phones. The quality looks different. They have higher resolution. In general, they make you look more motivated as a seller than the person who simply takes boring pictures with his phone.

If you feel you were able to get the perfect shot, there is still more to be done. You might need to edit the colors to enhance the image. A good photographer knows how to put his touch without making it so apparent that he did. That is a great skill; the ability to edit a picture such that it looks just like the original thing but with a little extra neatness to it.

As I'm sure you have realized by now, there is a lot that goes into professional shots than meets the eye (once again, pun unintended). These things that I have mentioned apply to general shots. You are taking pictures of the property, and that is architectural photography. It is different from other types of pictures. There is specialized knowledge that you are going to want to have access to.

So why the long lesson in photography? Very simple. As capable as you might feel, you need to realize you're likely unqualified to handle this apparently simple task at a high level. And so is your listing agent. You're going to want a professional photographer who has a trained eye and the experience bring out the best in your home. So, you should invest in a good listing agent who will subsequently invest in a good photographer.

More statistics for you, homes with high-quality photography sell **32 percent** faster than homes without.

Homes that have more pictures also sell faster. A home with only one picture spends an average of 70 days on the market. A home with 20 pictures spends 32 days on the market. What does this mean?

This means that, according to statistics, showing more than one photo of your house could cut the time it would spend in the market by half.

On top of that, a little bit of information that might be satisfying to know is that homes with high-quality pictures sell for higher prices. Homes in the $200,000 to $1 million range tend to sell for $3,000 to $11,000 more.

Take advantage of an expert photographer's knowledge to save yourself time on the market and make yourself more money.

CHAPTER 7

HOME STAGING

We all watch television. Real estate shows have been a popular part of everyday life for a long while. You must have seen a movie or two where an agent is hosting an open house. The house is filled up with furniture; everything looks nice and beautiful there might even be snacks. People are coming in and out, with smiles on their faces. They are ready to buy a house!

Perhaps it is just one person being shown around. The real estate agent leads them into all the rooms. They are talking. Probably even discussing how to turn this room into that and that room into this. The home looks furnished enough and ready to go. Those houses were staged.

Home staging is preparing a house for sale. You are taking an empty house and furnishing it. The goal is to make it look more appealing and attractive to the buyer.

It is like wrapping a gift except that this gift comes with sofas, chairs, tables, cushions and whatnot instead of wrapping paper.

It is not enough to **just** clean a house. Although, you could do that and be done with it. An empty house is like a blank canvas; a tabula rasa. The possibilities are endless, and the thought of that is exciting both for you, as the seller, and for the buyer. With home staging, you can show them the possibilities.

Home staging enables potential buyers to look and see what exactly they can do with the space. They are staring at a couch and thinking, "I will get a different color, I'd move it from there to there." You might not be privy to all these, but your buyers would begin to visualize living there. Once they start to do this, they are one step away from buying your home.

Ever entered homes that felt cold and impersonal? Perhaps it didn't feel cozy enough; minimalist architecture taken too far. The point of home staging is to make the buyer feel at home. There is a catch. You need to do this while also making sure the furnishings to the home are not personalized. You want it impersonal, but you also want it cozy. You want it furnished, but you don't want clutter either. When your buyer enters the property and begins to feel at home, it won't take much more for him or her to decide to make it **their** home.

There is more to staging a home than just filling it with furniture. I already mentioned that you need to depersonalize the whole space. You don't want them walking in and seeing trophies from the high school spelling bee you took first place in. Take down all those pictures and memories. It's a distraction to potential home buyers. You need to take that home and make it pop!

You should aim at making it look like a movie set, or perhaps a hotel room. Desensitized yet very welcoming. If you've ever viewed a new construction model home, you may have noticed you absolutely pictured yourself living there. And that was entirely by design. For that reason, you want the styling to portray the home as a very low maintenance living space. Even if it's far from what your home looks like on the day to day basis. Truth is, nobody's home looks like a model home. People have families, families are disorganized. But you don't need to reveal that yours is too. You want the home to be doing the talking and convincing, allow your home to catch its buyer and not the other way around.

Real estate research indicates about **98%** of buyers lack the ability to visualize how to use empty spaces. 98% is a really large number if you ask me. That's pretty much everyone.

This is not just speculation. Experts say that well-staged homes sell faster and for an average of 6% above asking price. A potential six percent raise would be enough to help me understand the value of home staging.

CHAPTER 8

PROFESSIONAL VIDEO

Real Estate Video

We have spent a whole bunch of time explaining just how beneficial it would be to seek professional pictures for the home. It is just the same with videos. It might be even more so. You want your buyers to feel something from watching your content. You want them staring at their screens thinking, "I want to visit this home."

Real estate listings that include video tours receive far more inquiries than their counterparts. In fact, a recent survey revealed most buyers and sellers prefer to work with agents that understand the usefulness of making real estate videos for marketing.

There is a whole demographic out there waiting for you to reach them. After doing all that is expected in preparing a home, taking that extra step to fulfil this group's love for video would work wonders for your traffic and eventual sale.

With people, so many things are determined by optics. Humans tend to pay less attention to other details if an item looks good. You are showing them your home, but you are also giving them everything they want to see and more. It will work.

Concerning the sort of videos, drone shots are also a big thing. This is because drones are becoming very popular now. Not only for videos but for pictures too. Everyone can appreciate the vantage point an aerial view of the neighborhood their future home is in when shopping around.

Still in the ballpark of professional videos, **3D virtual tours are now a must.**

Real estate listings that are accompanied with virtual tours enjoy a significant increase in the number of views they get. To put it in perspective, the number of views are inflated to a rate of up to 80% more.

Since COVID-19, you have not seen the same freedom as before to check out houses. Guidelines and regulations made it more difficult. You can't just up and say you want to drive down and tour a house. There is the issue of safety to think about now.

Virtual tours for real estate have become so imperative in real estate marketing to avoid unnecessary foot traffic through a home, especially after COVID. It's no surprise listings with virtual tours typically spark more interest than those without virtual imaging.

Consider customers from out of state who are looking to buy a home. A Real estate virtual tour is like a round-the-clock open house. It still gets the job done even when interested clients can't be there in person to tour a house. They can do it right from the comfort of their homes.

Even before the pandemic and movement restrictions, there would be a customer or two, who, for some reason, find it challenging to make it down to check out a home. Real estate virtual tours make it easy to solve this problem.

Virtual tours keep people on your website for much longer.

Planet Home's "Trend Study" concluded that, "75% of potential customers and visitors consider a virtual tour to be a major decision-making tool before proceeding to buy a house." Do you hear that? Major decision-making tool. So, they see a house they like and check to see if it has a virtual tour and if it doesn't, they move on to the next one. Imagine doing all the work to catch their eye, pique their interest and get them there, only to lose them because the buyers ultimately spent more time looking your competitor's home, which had virtual tour. Virtual tours are now a curial part of marketing; it has become very clear.

Property Week has also found out that, "virtual tours reduce the number of wasted viewings by 40%." This means that if buyers are surfing around on a website, they are less likely to waste time scheduling an in-person showing after their 3D virtual walkthrough assured them the home was not what they're looking for in the first place. How would you like to reduce the number of showings to people who weren't even going to like your house in the first place? Virtual tours can not only save the buyer time, but they can also save the seller from having to prepare for a showing that was not bound to be a good fit for the buyer. As a seller, you're saving yourself not only time but energy. You're cutting down the number of unnecessary showings. And buyers would be able to make their decisions easier, faster, and with more certainty.

How much more time do you think a buyer would spend on your listing if you added a 3D virtual tour? Your buyers would be able to explore your home for as long as they need to. They would be able to picture themselves living there and familiarize themselves with the layout. The importance of this can't be overrated. You don't want your house to be just like every other house on sale. First impressions are long lasting and cant be undervalued. In general, people don't remember what they see as much as we remember how what we saw made us feel. The feeling a first impression creates is what will, as they say, "last forever." You want the long lasting first impression of your home to cause the buyer to associate your home with comfort and facilitation. Not to be associated with the negative feeling of a homebuyer's difficult experience in trying to guess the layout of your home. That may lead to frustration.

When a homebuyer associates your home with gratification, don't you think they will remember your house more than the rest? The best part is **95%** of home buyers are more likely to call to inquire about a home if the listing includes a virtual tour versus a listing that doesn't.

CHAPTER 9

THE HOMEBUYER PSYCHOLOGY

When you can understand a person's motivation, convincing them of what they already want becomes a whole lot easier. Often time, you would hear people referring to a sale as meeting a customer need: giving the solution to the problem that a customer needs solving. What about a situation where your answer is just one out of multiple alternatives, what do you do then?

With businesses, a major goal is to understand the customers. This is more challenging than it seems. Let's say you are able to figure out a particular set of people. There are still a bunch of others with different interests and preferences. Well, what about them?

Excited about another psychology lecture? Don't worry. It is not going to be something boring about neurons, stimuli, and impulses. It can even be seen as the opposite. A branch of psychology that talks about understanding your buyers, and the science of the sale is behavioral economics. As the title suggests, it is about understanding the behaviors behind activities of commerce. Let's call it the psychology of the "buy" or buyer's psychology.

There are different things that your customer considers, either consciously or unconsciously, before making a purchase. The first is their need. A man won't just wake up one morning and say he wants to buy a property. He needs it for something. It could be as a place of residence. It could be for rental income. It could be for anything, really. There is always a motivation behind it which can spark their enthusiasm to take action; that is the takeaway here.

Buyers are also led by how they feel. Human beings are emotional creatures. We place a premium on gut feelings, intuitions, and feelings in general. It is in your best interest to influence their emotions in your favor. It does not matter how good a person is at rationalizing; a lot of purchase decisions are tied to emotions. It might seem rational and well thought out when they explain, but that might just be a nice packaging for the real reasons. Having established that buyers are emotional; you want to tap into this characteristic by developing the marketing of your home to appeal to their emotional side. And make them feel at home. You could achieve this with the home staging, pictures, videos or 3D tour. You could achieve it with **all** of them.

This section is the fun part of selling your home; understanding the psychology of buying. It feels good to be able to understand your customers, and what gets them excited. Apart from how it feels, it is needed if you want to tap into the trigger that causes homebuyers to pay more for your home.

You might want to believe it and you might not, but it relates specifically to the photos and videos of your home. Yes, we are back to the optics once again.

Have you ever wondered what home buyers are going to feel what they look at your house? What are they supposed to feel? This is very vital. In that split second it takes to form a first impression, you want to have been able to communicate your most valuable message to your buyers. After they have reached that impression, you are going to want to reinforce it.

Home sellers have to understand that marketing a home is not as simple as just uploading nice pictures of the house on the internet and then packing up shop. Marketing is essentially creative storytelling. A story that subliminally highlighting the home buyer's needs. I already told you how much your buyers love a good story.

Stories make the buyers feel some type of special way emotionally and socially. A story can make an impact through words and visuals, while building a connection and leading the buyer right into your home. This is why a good marketing campaign could be a home seller's perfect opportunity to engage with his buyer's brain. It's the marketer's job to figure out how to influence the trigger in your home buyer's brain to lead them to feel they absolutely want your home and make an offer to buy. We all have that little moment when things just click, when the decision is made that, "I need to buy this." Your marketer's goal should be to induce that little moment. The trigger is the emotions—the reason why (once again) is that emotions drive behavior.

We buy a lot of things because they make us feel good. Humans are emotional buyers. You don't think this is true? Ask yourself a simple question. Have you ever bought something you don't need?

If you answered yes, which I'm positive that you did, that means you're an emotional buyer, or you might have been an emotional buyer at a point. You should be looking at how to make your buyer feel good about the house. Your job as a home seller is to contribute in bringing out those emotions from a buyer when they look at your home. Maybe even trigger the dopamine surge that can lead to an action to buy.

The level of emotion we feel and how we remember something is directly related to the amount of dopamine the particular thing we see, or experience, is able to release. This is very similar to the reason a lot of people get addicted to social media and video content is a major part of it.

Did you know that videos are the most engaging content across all social media? According to Business Insider, which we all love and trust, video will represent 82% of all internet traffic by 2021.

Doesn't that sound like a useful piece of information for someone who wants to bring attention to their home? You can use it to your advantage.

People tend to like videos more than they do pictures. Even on social media. In my opinion, the reason videos gain more traction is because of the music. A well-edited video coupled with pleasing music is going to be vital in capturing the emotions that you want and triggering the dopamine increase that will make home buyers remember your home in a positive manner. This is because music influences people's drive. Any type of music that you are listening to has an impact on your level of motivation. It changes the way we feel. Homebuyers feel the same way. They might not realize it, but their minds take note of the music and react accordingly.

Music

Music has the ability to evoke strong emotional responses. Ever felt so much energy listening to an upbeat song that you just get up and start dancing? Research on music reveals pleasurable music may lead to the release of neurotransmitters associated with reward, such as dopamine. The brain is wired to reward things that make us feel good so that we can have more of that. And we know that listening to music is an easy way to alter mood or relieve stress.

If anything, remember one thing. Music creates emotion. Emotion leads to action. The power of music in the video marketing of your home could subconsciously trigger just the right amount of emotions in homebuyers to influence them to rank your home just a tad bit higher than your competition.

This is your perfect opportunity to get surgical with marketing to take advantage.

Most importantly, you need to understand this. You are not just selling real estate. You're not selling four walls and what's inside. You're selling the lifestyle that comes with it. You are selling a way of life. It is a package deal. The shell of the house is just one of everything.

Through your marketing, buyers should be reminded of why it is they would like to live in your neighborhood in the first place. You're beginning to see elements of the psychology of buying play out here. The story in the marketing of your home should remind the buyer if the home is close to the beach. Make it a point to tell them about the pool. If they are a family, take your time to tell them about the schools. Let them know about the highly rated schools in the area. Tell them about nearby shopping centers. Note the parks, the shopping centers, or the good restaurants in the area. Let them know about all the other things of interest in the area apart from the house. That's what the story is. It's the unwritten story of their lives in their future home. Your current home.

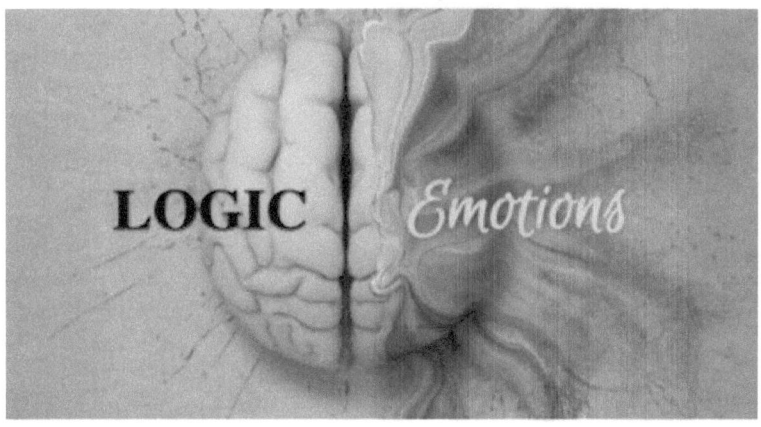

Remember: People don't just buy things; they buy feelings. We tend to purchase the things that trigger a dopamine release. Once again, the brain is wired to reward things that leave us with a positive feeling. It reinforces it so that we can have more of it. The way we remember something is directly associated with dopamine. Fond memories equal higher levels of dopamine and vice versa. We buy things based on our emotions, and we justify it with logic. Sound familiar?

The cocktail of the right pictures, videos, tours, proper home staging, and information about your market is a master brew that would help you to package your home as the best alternative out of all the other options in the market. The chances are that the competition might drop the ball in one area or the other. You need to remain on top of your game at all sides in order to separate your home from the rest.

CHAPTER 10

PRICING

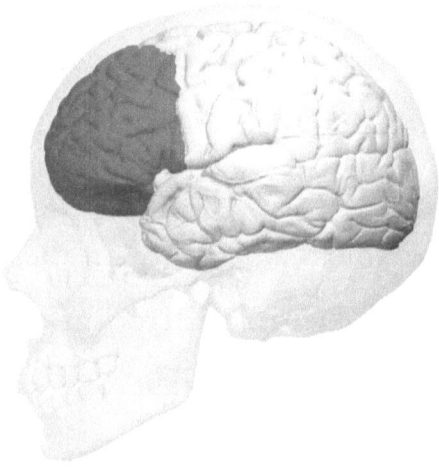

This is an area where a mistake could cost sellers a lot of money. Poor pricing causes home sellers to sacrifice their equity. But sometimes, sellers make poor pricing choices due to poor advice by real estate agents.

It's common knowledge that homeowners usually think very highly of their homes. That is completely normal and also expected. We all do. We believe our house is the best one in the neighborhood, just like we adore our kids more than any other child out there. This view of our homes is entirely unbiased, right?

Homeowners are attached. You are going to first discover this at the point of home staging. It is a home you have lived in and have gotten used to. The process of decluttering might be challenging. You're getting rid of every form of intimacy you share with the home, save for your name being on the deed. Learn to detach yourself. Depersonalize the home and get comfortable with the fact that you will be trading ownership.

Experienced real estate agents know this. They've been in the game long enough to know that if it were up to you, you would very likely overprice your home considerably. They are also aware of the fact that one of the fastest ways to lose a client is by being honest about the home's value and disappoint the would-be sellers. No seller would like to hear an agent come up to tell them that the home they so much love and appreciate is actually worth less than they think it is. Thanks, but no thanks. So, one of the oldest tricks in the book is for a real estate agent to tell the seller that their home is worth more than it actually is. They feed into the mentality that you already have, and they do this for very obvious reasons. They want you to list with them.

You have emotions too. To the agent, you're the buyer buying their services, so they are going to want to influence your feelings in the easiest way possible. When you hear the high price valuation on your home, you're going to want to act on it. That's not the way to look at it. You would end up taking a deep breath, closing your eyes and shooting yourself in the foot. When you hear that high price, it is probably going to tamper with your emotions. You don't want that to happen; you want to lead with a clear head. If you let it mess with your feelings, you're probably going to end up listing for the wrong reasons. Home sellers should be very cautious of that.

In the long run, allowing yourself to be emotional about the pricing decision would be very costly. Generally, after a week or two, the listing agent's solution to the overpriced listing would be to go to the seller and request a price adjustment, knowing very well that the property was overpriced, to begin with. That is a great disservice to the seller.

If your home is priced too high from the start, logic is going to trump all emotion. Customers don't want to make a loss. This causes your home to sit on the market for longer than expected because nobody wants to make an offer. When your home is sitting on the market for way too long, people begin to wonder why. They start to wonder why it's spending so long, and if there is something wrong with your home.

Thinking your home may have issues that are keeping it from selling, buyers will make unnaturally lower offers or don't offer at all.

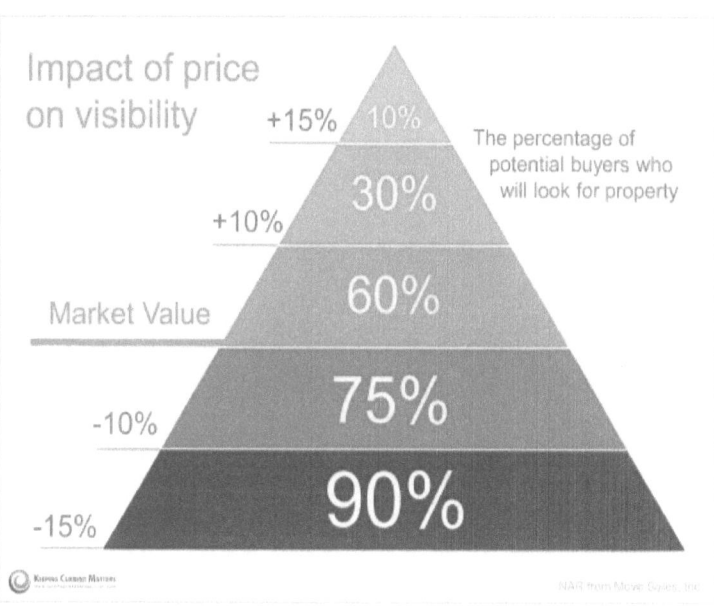

From the diagram above, you can see that price has a massive impact on visibility. The percentage of buyers that look at property that is priced at up to 15% above market value is as low as 10%. As the price reduces, the percentage of potential buyers also increases. You might lose your best buyers if you price too high.

You want to avoid mistakes with pricing at all costs because, typically, the longer a house stays on the market, the lower that house is going to sell for.

It is essential for you to understand that. Here's why:

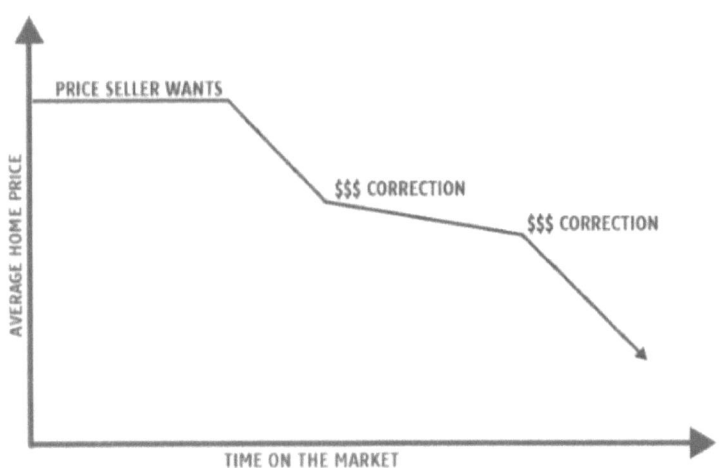

If you screw up from the start, price corrections become inevitable. You would be forced to make adjustments. Poor pricing will cost you money. You would be sacrificing your equity as a result of the poor decision. Losses like that make it harder for you to read your initial goal, which is to sell for as high as possible.

It is like spoiling your asking price. It is no longer any good. You have to throw it away and replace it with a new one; a way more attractive one. It is no different from a poorly priced perishable product at a grocery store. Poorly/overpriced items are commonly passed up by the consumer. When multiple products of the same use are priced differently, people are used to selecting the cheaper of the bunch. We all do this. It's normal, and it shouldn't come as a surprise.

Believe it or not, unless your home absolutely stands out from the rest, a home buyer will be inclined to go for the cheaper option. This causes your home to stay on the market longer, forcing you as a seller to drop the price. The graph above appropriately depicts this. The high price leads to price corrections which eventually leads to the home being on the market for a longer time.

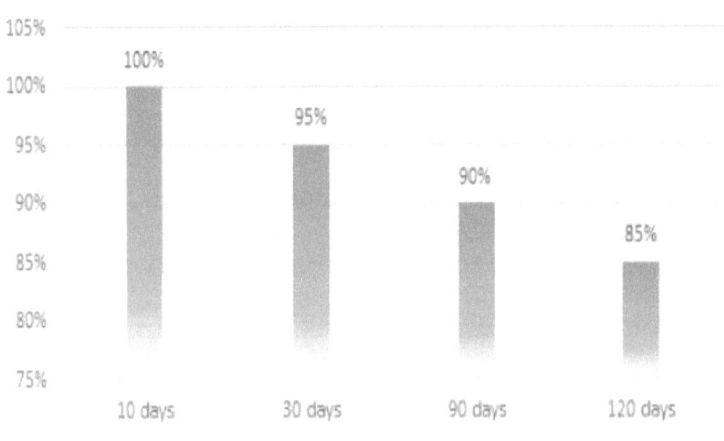

The amount of time that your house spends on the market matters significantly. On average, if you sell within the first ten days, you are very likely to sell for 100% of your asking price. If you sell within the first 30 days, you can expect about 95% of your asking price. If you sell within the first 90 days, you are likely to sell for 90% of your asking price and so on. This proves that those beginning stages of selling a home are crucial. You already understand that the longer a house spends on the market, the higher the chances of sacrificing equity because the buyers are beginning to wonder why it's spending so long on the market. You are also forced to be making those unappealing price adjustments.

Now let's talk about the realm of market prices and the popular estimate of the value of your home. There is a pre-determined price for your house. That price is the highest that your home could sell for. I don't know the number, you don't either. Even a psychic doesn't know the number.

What you **do** know is that similar houses are being sold. Buyers are paying based on market research. In a healthy market, recent sales comps (Comparables) give us a general idea of what your home's market value is.

Suppose you have a crystal ball, a crystal ball that actually works. Most of us would use it to predict the winning lottery numbers or something like that. But in this scenario, let us imagine that the crystal ball tells us that the maximum amount a homebuyer would be willing to pay for your house is $500,000 and not a penny more. Now this tells us that there is a buyer out there. There is a buyer who is willing to pay five hundred thousand dollars for your home. We don't know who this home buyer is, but we know that he is out there among the rest.

Let's make another assumption and say that you are like most homeowners, and you thought your home was worth more than that. You wanted to sell it for higher. For the purpose of this example, we will exaggerate and say you think your home is worth $1,000,000. Because of this, you decide to get your home listed at a million dollars. Even though you overpriced your home, that interested buyer still does his research on the market value and quickly reaches out. He makes his highest, and best, offer to you. How much do you think your home would sell for, assuming you accept that offer?

The right answer is this. It is still going to sell for five hundred thousand dollars. The maximum amount is **still** the maximum amount.

What if you were to price it ridiculously low? In this case, let's say you price your beloved home at a hundred thousand dollars. It is ridiculously underpriced, but I'm trying to prove a point. Humor me.

The answer is still the same: five hundred thousand dollars. The maximum amount a person would be willing to pay for that house would still be the same. It' wouldn't change.

It wouldn't change, and it would remain at that pre-determined price unless you are able to generate enough homebuyer traffic that could lead to the production of multiple offers. This is where an experience listing agent would be able to leverage your position, negotiate the offers, and create a bidding war. Sounds exciting, doesn't it? It **is** exciting but note that you will **not** create a bidding war if you overpriced your home.

However, if you reasonably underprice your home and you trust your marketing plan, you can put together a suitable pricing strategy with your listing agent. Take note of the word, **reasonably**. The above example was an exaggeration. Although, I have seen it done in real life

You can create a bidding war for your home, no matter the location. Here is why. It is called FOMO; The Fear Of Missing Out.

You know about how people feel a need to **not** be behind the curve. They want to be up to date with trends, and what is in style. This also causes them to have a feeling of FOMO. The fear of missing out is an anxiety people experience when something exciting is going to take place in their absence. You've most likely experienced it whether or not you're aware. That anxiety can be leveraged with home sales too. If you market your home correctly, buyers might feel like they are missing out on a great opportunity if they don't make their highest and best offer quickly enough.

Once a person gets wind that a home is in their best interests and it is a great bargain, they tend to latch on. The realization that other people are doing the same makes them latch on with a firmer grip. The other parties are also feeling the same way and doing everything they can to hold on to the deal. It becomes more than just another house on the market that they are considering. It becomes a pretty package that would slip through their fingers if they let it. I don't know if you have noticed this, but people tend to spend more when they feel like they are getting a great deal than they would spend otherwise. Think back to a time you, or someone you know, heard of a discount that was going to end soon. That person probably bought more than they usually would. It was a great bargain, right? They didn't want to lose the opportunity.

As home sellers, we need to do the same in order to create a bidding war. Urgency can really motivate buyers to "hurry up" and buy your home before somebody else does. This is going to be very good for you.

One of the first things a buyer looks at when searching for a home is the price. If the home is well priced, they will continue to show an interest in it. And if you effectively satisfy the home buyer's needs and you market your home appropriately, buyers are going to fall in love with your home. Their emotions would be higher than that of a person looking at an overpriced home.

Logic and emotion continued...

When people like a home and they believe the home is reasonably priced, it is normal for their feelings to naturally gravitate towards that home. Their interest in it would spike. An experienced real estate agent representing the buyer would most likely recognize the fact that your home is underpriced, and would alert the buyers of the opportunity, and the importance of taking fast action. You've succeeded in creating some urgency now.

Logic tells them that if they see it as a good deal, other buyers would too. They would hasten up and rush to the bargain, so they don't lose the deal. They are all going to want to be the first to make an offer and get the home, and that is precisely how bidding wars are started.

Think about it. The homebuyer doesn't want to miss out. They are thinking about the other buyers and wondering about how much they would be offering. They already think of the house as the real deal. Their next move is to offer higher. It is now their intention to make the most attractive offer. Chances are, they would offer even higher than they had initially intended to spend on the house.

My opinion as a home seller is that you position yourself in the market to appeal to as many buyers as possible. The point is to get the house sold as quickly as you can so that, as a seller, you don't have to live through the anguish of worrying about the potential of the home not selling, and potentially being forced to reduce the price, especially if you're in a downhill market and are playing catch up with the pricing.

It is better to discount your recent sales and comparables ahead of time than to have to adjust the price over and over. You already know what happens if you are forced to do that. It is kind of like shooting at a moving target. You don't shoot where the target is, but rather you anticipate and aim at where you think the target is going to be.

Remember, don't let yourself be influenced by a listing agent suggesting an unrealistically high price to you. You should have a very good idea of what your home is worth prior to deciding to go on the market. It is your house that is going up for sale, not your real estate agent's. Even though you should trust an expert's advice, it is always good to do your own research so that you can trust your gut feeling.

CONCLUSION

At this point in the book and in time, Tom, Jack and Jill, The Smiths, and the bachelor are pretty much armed to the teeth with useful information on how to sell their homes. From understanding the principles of demand and supply to how they affect pricing, and from learning about home buyer traffic to understanding their market, they are well equipped. All of that coupled with the information about useful data, and the impact of social media with the use of effective homebuyer retargeting gives them an edge over their competitors who simply might have planned to hire an agent and be done with it.

That is all well and good, but they are also accompanied with knowledge about the importance of professional photography, home staging, professional videos and 3d tours. By now, they have no problem capturing and keeping their buyer's attention. They cater to home buyer psychology and give the buyers precisely what they need and want while triggering the right dopamine surges that facilitate an attachment to the home. They do all these while using a very well thought out pricing strategy.

This book promised to introduce innovative systems that would enable sellers like them to get the best deal and make minimal errors. Follow their footsteps and make use of everything that is now in your arsenal. Reduce the possibility of incurring losses and being misled by real estate agents who understand how the game works. Even with the 2020 pandemic causing a change in how things are done, you still have more than enough to leverage. Your 3D virtual tours are a capable replacement for unnecessary home visits. They can also convince buyers who were on the edge to actually schedule an in person showing instead of immediately cross your home off the list. Then there's also professional videos which aside from facilitating exposure, they would be a whole lot in appealing to your buyers' emotions.

Detach yourself from the home momentarily, think like a business owner, and leverage your home as an income producing asset. Seek to invest in driving large homebuyer traffic to your home and give those buyers the best experience they can wish for. It will pay dividends. Giving them what they want eventually means giving yourself what you want; more money. You don't want to drop the ball and deliver service that is less than top-tier. You are a businessman when it comes to this sale, and you are going to apply a businessman's clever tactics to enchant your potential buyers and get yourself better proceeds.

You intend to sell your home as fast as possible while making as much profit as possible. What are you waiting for? Grab some scissors and cut that thread of awareness we discussed. Good luck.

www.ingramcontent.com/pod-product-compliance
Lightning Source LLC
Chambersburg PA
CBHW022107160426
43198CB00008B/388